cience all around me

Electricity

Karen Bryant-Mole

Heinemann

First published in Great Britain by Heinemann Library, Halley Court, Jordan Hill, Oxford OX2 8EJ
a division of Reed Educational & Professional Publishing Ltd.

OXFORD FLORENCE PRAGUE MADRID ATHENS MELBOURNE AUCKLAND KUALA LUMPUR
SINGAPORE TOKYO IBADAN NAIROBI KAMPALA JOHANNESBURG GABORONE PORTSMOUTH
NH (USA) CHICAGO MEXICO CITY SAO PAULO

Designed by Jean Wheeler
Commissioned photography by Zul Mukhida
Consultant – Hazel Grice
Printed in Hong Kong

00 99 98
10 9 8 7 6 5 4 3 2

British Library Cataloguing in Publication Data

Bryant-Mole, Karen
  Electricity. - (Science all around me)
  1. Electricity - Juvenile literature
  I. Title
  537

ISBN 0 431 07825 4

A number of questions are posed in this book. They are designed
to consolidate children's understanding by encouraging further
exploration of the science in their everyday lives.

Words that appear in the text **in bold** can
be found in the glossary.

Acknowledgements
The Publishers would like to thank the following for permission to reproduce photographs. Chapel Studios 16; Eye Ubiquitous 6;
National Power 8; Positive Images 9, 12, 14; Tony Stone Images 4 (Paul Redman), 10 (Mary Kate Denny); Zefa 18, 20, 22.

Every effort has been made to contact copyright holders of any material reproduced in this book. Any omissions will be
rectified in subsequent printings if notice is given to the Publisher.

Contents

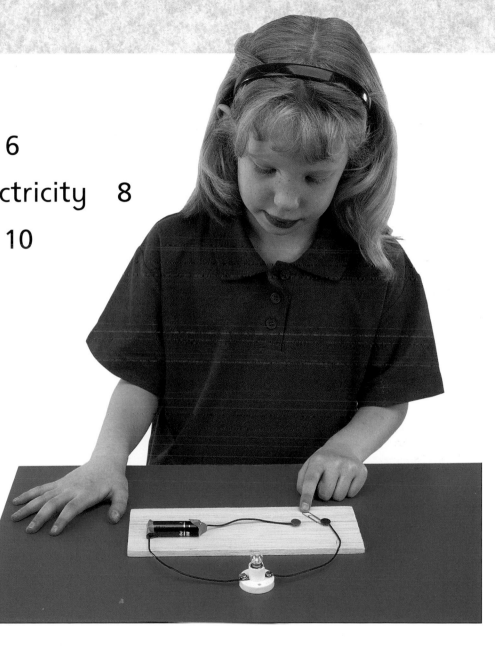

Electricity

Electricity is a form of energy.

Energy can be difficult to understand because it cannot usually be seen.

A way of describing energy would be to say that it makes things work.

Lots of the things that we use every day need electricity to make them work.

? Which of the things in this picture need electricity?

See for yourself ...

We often use electrical things in one particular room.

A toaster is usually found in a kitchen.

Leila has drawn a big picture of a house.

She has found some pictures of things that need electricity.

She has stuck them in the right rooms.

Static electricity

During a thunder storm, tiny water droplets in the clouds rub together and become charged with something called **static electricity**.

The electricity in the storm clouds then moves between the clouds or down to the earth.

It makes a very big spark.

? *What do we call this big spark?*

See for yourself ...

Tom has charged a plastic spoon with static electricity, by rubbing it very hard on his shirt.

Now he is holding it just above some salt.

Static electricity makes the salt jump up and cling to the bottom of the spoon.

Manufactured electricity

Electricity can be made, or manufactured, in a special building, called a power station.

This electricity is moved along power lines to homes and businesses.

Electricity that moves from one place to another is called current electricity.

?*Can you see some wide towers? They are called **cooling towers**.*

See for yourself ...

As electricity enters our homes, it passes through a **meter** that measures the amount we use.

Bartie's dad is showing him the electricity meter outside their home.

Bartie can see a dial spinning.

The more electricity they use, the faster the dial spins.

Using electricity

Electricity is one form of energy.
It can be changed into other forms of energy.

Electricity can be changed into heat energy by a toaster, light energy by a light bulb or sound energy by a radio.

(i) *Electricity can also be used to turn a **motor**. Dishwashers and vacuum cleaners have motors.*

See for yourself ...

Naheed is looking at some of the electrical equipment that is used in his home.

He is trying to decide what forms of energy the electrical energy is turned into.

Perhaps you could help him?

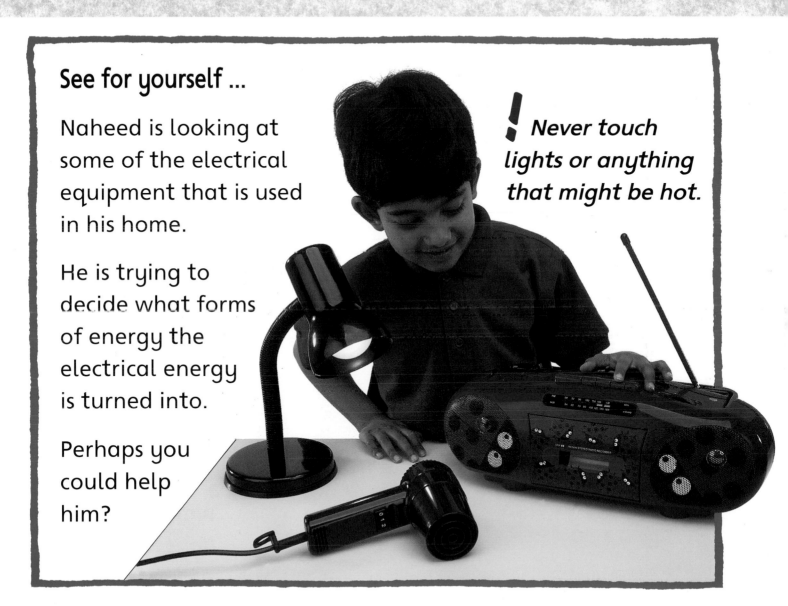

! *Never touch lights or anything that might be hot.*

DANGER

Electricity is very useful but it can also be very dangerous.

The electricity in your home is powerful enough to kill you.

Pylons that carry power lines across the countryside have warning signs on them.

! *Never play with plugs, light bulbs, sockets or wires.*

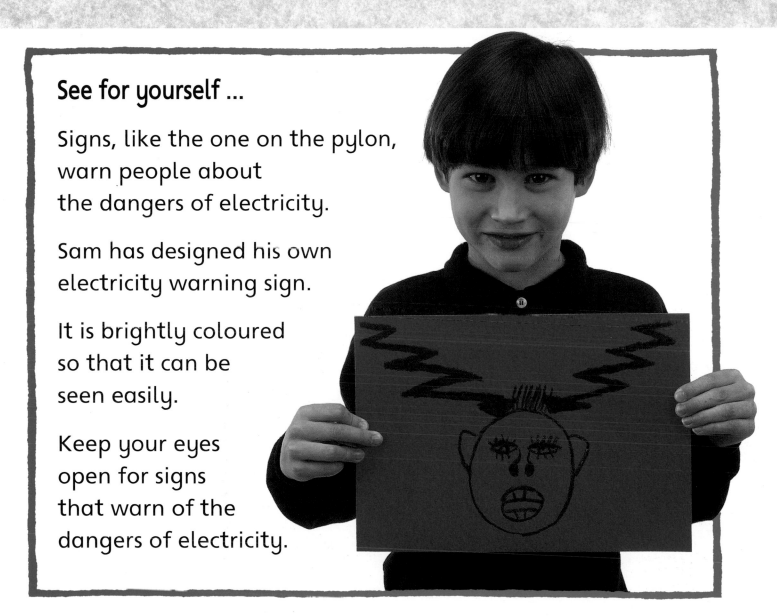

See for yourself ...

Signs, like the one on the pylon, warn people about the dangers of electricity.

Sam has designed his own electricity warning sign.

It is brightly coloured so that it can be seen easily.

Keep your eyes open for signs that warn of the dangers of electricity.

13

Batteries

This personal stereo is powered by batteries.

Batteries have chemicals inside them that can change and make electricity.

When all the chemicals have changed,
the battery stops making electricity.

? *Can you think of any other objects that can be powered by batteries?*

See for yourself ...

Robert is testing some batteries.
He puts them in a toy
that takes one battery.

If the toy does not work
when he switches it on,
all the chemicals in
the battery have changed.

Batteries like these are safe
to use because their electricity
is much weaker than
the electricity flowing
around our homes.

Circuits

The bulbs and wires in these Christmas tree lights form a circuit.

A circuit is a loop around which electricity can flow.

If there is a break in the circuit, the electricity will not be able to flow all the way round.

(i) *If one of these bulbs stops working, all the lights may go out.*

See for yourself ...

Naheed is using a battery, a bulb in a holder, two pieces of wire and some tape to make a circuit.

The bulb is not shining because electricity cannot yet flow all the way around.

If he joins the end of the wire to the battery, it will make a loop and the bulb will light up.

Switches

When you want to watch television or turn the lights on, you press a button or flick a switch.

Switches are used to **complete** or break a circuit.

Switches turn equipment on or off.

? *How is the electrical equipment in your home turned on and off?*

See for yourself ...

Alex has decided to make a circuit with a switch.

She has used three pieces of wire, two drawing pins, a paper clip and some balsa wood.

She can use her paper clip switch to complete or break the circuit.

This makes the bulb flash on and off.

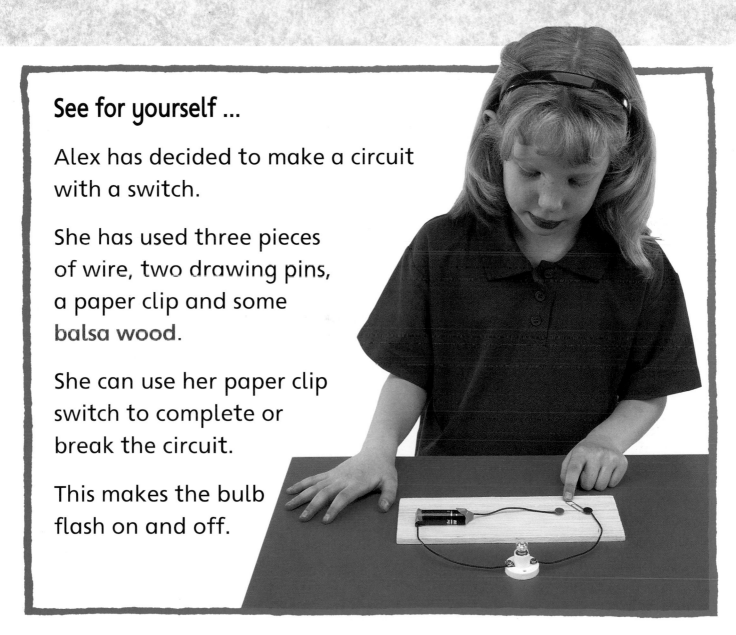

Conductors

Materials that let electricity flow through them very easily are called conductors.

Most metals are good conductors.

Trams use metal wires to conduct electricity from the overhead wires to the tram.

Our bodies conduct electricity, too. If electricity passed through someone's body, it would give it an electric shock.

! *Never play with electricity.*
An electric shock can kill.

See for yourself ...

Naheed wants to see a conductor working.

He has folded a piece of foil into a strip and made a gap in a circuit.

If he tapes the ends of the wires to the ends of the strip, the bulb will light up because foil conducts electricity.

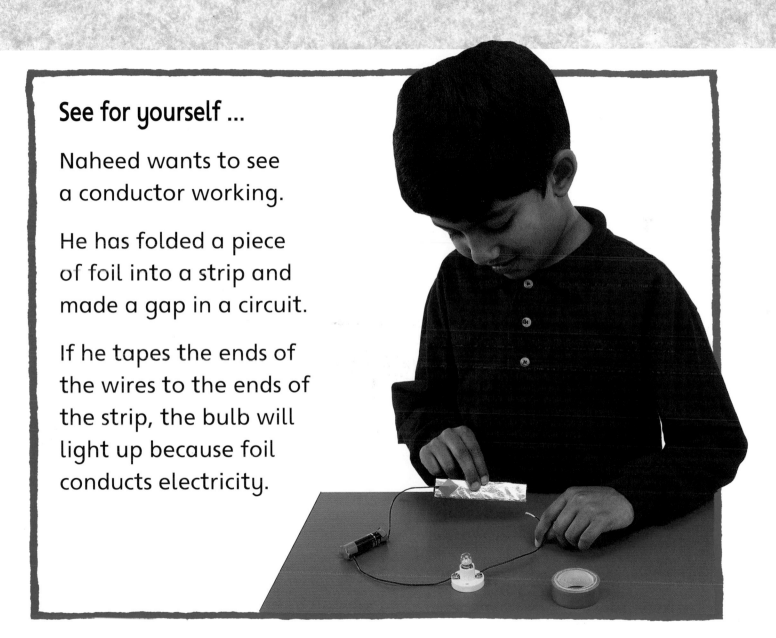

Insulators

Materials that do not let electricity flow through them are called insulators.

Most plastics are insulators.

This iron has a plastic case to protect the person using it.

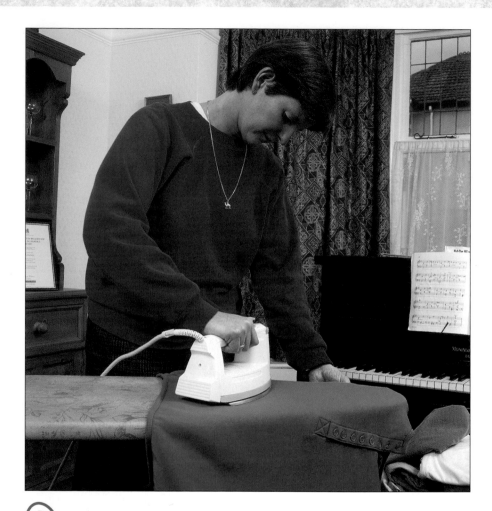

(i) *The wires that carry electricity in our homes are usually covered in plastic, too.*

See for yourself ...

Alex and Sam are finding out which materials are insulators and which are conductors.

They have found some objects and are using tape to connect them into the circuit.

If the bulb lights up, the material that the object is made from is a conductor. If not, the material is an insulator.

Glossary

balsa wood   a light type of wood

complete   finish, make whole

cooling towers   places where water that has been used in the power station is cooled down

meter   a machine that measures something

motor   an engine that makes things work

pylons   tall, metal towers used to carry power lines

sockets   the holes into which a plug is pushed

static electricity   electricity that stays in one place

Index